TELECOMMUNICATIONS

BY COLIN HYNSON

TELECOMMUNICATIONS – WHAT IS IT?

The ancient Greek word *tele* means far off or far away. Telecommunication is the ability to communicate over large distances. In our modern world this is almost always in an electronic form through the telegraph, telephone, fax, radio, television, and the Internet. The form of the message has to be changed between the sender and the receiver. A message is usually converted into an electronic signal when it is sent off. When it reaches its final destination, the signal is turned back into its original form. The late 19th century saw the beginning of this new era of telecommunications when messages over long distances could travel at the speed of light. The 20th century continued with this revolution in long-distance communication. Now that we have just entered the third millennium, changes in telecommunications will become even better and faster. In 1900, there were no radios or televisions. The telephone was still in its infancy and was only used by a small group of people. What new ways of telecommunication will emerge in the next one hundred years?

RIDING ON WAVES

This picture shows a transmission tower that sends out radio waves. In order to communicate, waves on the electromagnetic spectrum are used. Radio and television use radio waves. For satellite communication, it is becoming more common to use microwaves. Even light waves are used in optical fibers to transmit telephone conversations.

LOOKING AND LISTENING

One of the greatest recent innovations in telecommunications is the development of the Internet. Telecommunications in the past has always been restricted by the kind of messages that it can send and receive. The Internet allows people to communicate instantly using words, sound, still pictures, animation, and video. The Internet is the television, radio, and telephone all rolled into one.

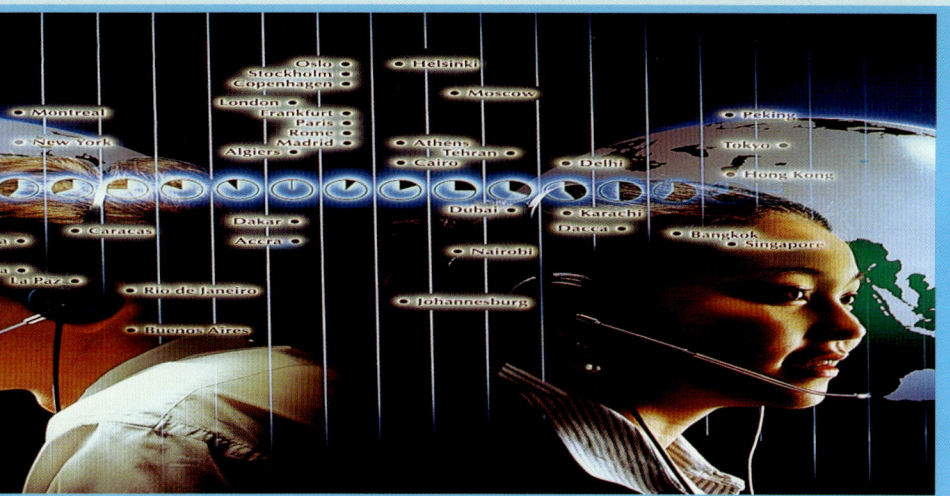

MESSAGES ACROSS THE WORLD

Modern telecommunications means worldwide communication. It is now as easy to talk to someone living on another continent as it is to talk to a next-door neighbor. Political and business leaders across the world can now easily contact each other to discuss issues that affect us all.

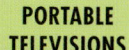

MOBILE TELEPHONES

The discovery of radio waves liberated us from the need to rely on cables to communicate. Today, mobile telephones enable people to send and receive messages no matter where they are.

PORTABLE TELEVISIONS

Developments in telecommunications have benefited from changes in other technologies. The introduction of the silicon chip to replace valves meant that radios and televisions could become much smaller, making them genuinely portable. Now people need never miss their favorite programs.

THE ARRIVAL OF THE CAMERAMAN

Instant and worldwide telecommunications mean that we are probably more familiar with what is happening in another part of the world than with events in our own town. In the past, news traveled extremely slowly, but today, major events all over the world are beamed live into our homes.

EARLY COMMUNICATION

For much of the history of the human race, our ability to communicate with each other has been limited by either speed or distance. Our voices travel at the speed of sound, which is about 1,086 feet (331 meters) per second. This is very fast indeed, but sound cannot travel very far through the air, so other ways of communicating are necessary. In the past, people who wanted to communicate across any distance were restricted by the speed of whatever was carrying the message, such as a horse or a ship. When the American colonies announced their independence in 1776, it took 48 days for the news to cross the Atlantic. It was possible to send visual signals, like smoke signals, which could be seen over long distances quite quickly. The problem was that whoever was trying to interpret them needed to know what the signals meant beforehand, otherwise they were meaningless.

THE PONY EXPRESS

The Pony Express lasted from just 1860 to 1861 and provided a rapid mail delivery service between Missouri and California. Messages traveled the 2,000 miles (3,200 km) in just 10 days, with 190 stations along the way where riders could change horses. The service ended when an overland telegraph system was successfully completed.

SAFETY SIGNAL

Ships coming into New York Bay in the 1830s could check the position of the arms on this pole to make sure they could sail safely in and out of the Bay. Signaling arms were also used at the side of railway lines, and some trains continue to receive signals in the same way even today.

BREAKING THE NEWS

One of the earliest messages carried by the Pony Express was this letter, which announced the election of Abraham Lincoln as President of the United States in November 1860.

A WARNING LIGHT

Lighthouses have been a source of comfort and information for generations of sailors. The light they emit can be seen for many miles, providing a navigational reference point and warning of danger like hidden rocks. The first recorded lighthouse was the great Pharos of Alexandria, one of the Seven Wonders of the World. Built in around 280 B.C., it measured about 350 feet (106 meters) high and had a wood fire at the top.

FIRE AS A SIGNAL

Probably the oldest method for sending a prearranged signal is by fire, as it can be seen over long distances. The Romans used this method, as did many civilizations before and after them.

MESSAGES THROUGH THE AIR

It was the Sumer people of northern Iraq who discovered, around 4,000 years ago, that pigeons always return to their nest. These birds were first used to carry messages for military purposes from about 1200 B.C. in Ancient Egypt. They were still used for carrying military messages during World War I, as this picture shows.

FLAG ALPHABET

In the past, ships communicated by the use of flags. Each flag stood for a different letter or number. The first recorded use of flags for sending messages at sea was by the Spanish in the 16th century. In 1799, the British Royal Navy standardized the flag designs so that all their ships could signal each other. The international code seen here was created in 1934.

5

OPTICAL TELEGRAPHY

In the years after 1789, the kings and queens of Europe watched in horror as a revolution in France overthrew the monarchy, executed both King Louis XVI and his wife, Marie Antoinette, and established a republic. European monarchs were determined that the French Revolution should be crushed, or they might face a similar fate at the hands of their people. By 1792, the new French government found itself at war with Austria, Prussia, England, and Spain. In the next year, German and Italian kings also joined the war. If the revolution was to survive against such odds, then a more efficient system of communication across the country was essential. The answer came in 1794 from Claude Chappe, who invented what he called a telegraph. The name came from the Greek word *tele* meaning distance and *graphien* meaning writing. It was the first ever national system of long-distance communication.

CHAPPE'S EXPERIMENTS

Claude Chappe was born in 1763 and trained to be a priest. He used the money he earned as a priest to buy equipment for scientific experiments. He was especially interested in the properties of sound and of static electricity.

HOW THE CHAPPE TELEGRAPH WORKED

The arms on top of the Chappe telegraph tower could be moved in many different positions. Each of them represented either an individual letter or number. This cut-away picture shows how the arms on top of the telegraph tower were actually moved. The main pole was about 20 feet (6 meters) high and was crossed by a bar with two movable arms at the ends. These could be moved into the required position by ropes and pulleys that were worked at the bottom of the pole.

STATIONED ACROSS THE COUNTRY

This new system of communication was so efficient that it could carry a message the 140 miles (225 km) from Lille on the Belgian border to Paris in about two minutes. A series of towers was built across much of France from 1794. They were usually on hilltops and were 5 to 10 miles (8 to 16 km) apart. A telescope was located at each tower so that the signals could be read more easily. By 1805, there were almost 1,000 telegraph towers throughout France.

MAKE A SIMPLE CHAPPE TELEGRAPH

To make your own working Chappe Telegraph you will need:

string, plastic modeling clay, a large yogurt container, three paper fasteners, tape or glue, stiff cardboard

1. Cut out one piece of cardboard measuring 12 in. by 1 in. (30.5 cm by 2.5 cm) and two more pieces that are 6 in. by 1 in. (15 cm by 2.5 cm) each.
2. Take the larger piece of cardboard and cut a small hole near the top. Now take one of the smaller pieces of cardboard and cut a small hole in the middle and a hole near each end (figure 1).
3. Line up the middle hole with the hole in the large piece of cardboard. Join them together using a paper fastener, but keep it loose enough to let the pieces of cardboard move freely (figure 2).
4. Now take the remaining piece of cardboard and cut it exactly in half (figure 3). Make a hole about 1 in. (2.5 cm) away from the end of each piece. Now attach these to the two remaining holes on your "telegraph" using paper fasteners, again not too tightly.
5. Add a small piece of clay on the bottom of these two smaller arms to add a little weight.
6. Cut four pieces of string, each about 5 in. (13 cm) long. Use the tape or glue to attach one piece of string to each end of the two arms with clay weights (figure 4). Stick the other two pieces of string on the 6 in. (15 cm) piece of cardboard about ½ in. (1.25 cm) on either side of the longest piece of cardboard.
7. Lastly, cut a 1 in. (2.5 cm) slit in the top of the yogurt container and push through the large piece of cardboard.
8. You can now represent the words "CHAPPE TELEGRAPH" on your model, using the table of arm positions below.

figure 1

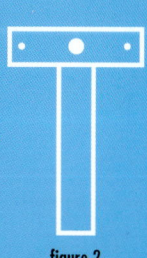

figure 2

figure 3

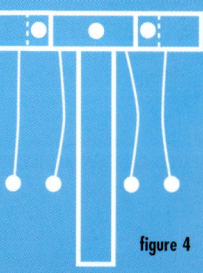

figure 4

FIGHTING FOR FREEDOM

The French Revolution began as a protest about high taxes. However, the storming of the Bastille prison in Paris in July 1789 led to a nationwide revolt. In August, the National Assembly issued the "Declaration of the Rights of Man and of the Citizen" that demanded rights for all. This started the revolution and created a new government based on the ideas of liberty and equality.

A	C	E	G	H	L	P	R	T

7

THE POWER TO COMMUNICATE

Knowledge of the existence of magnetism and electricity can be traced back to ancient times. The Chinese are said to have been using magnets to guide troops through fog as long ago as 2500 B.C. Around 600 B.C., a Greek philosopher named Thales noticed that if a piece of amber was rubbed with a cloth it attracted bits of fluff. This strange attractive power became known as electricity. During a lecture in 1819, the Danish scientist Hans Christian Oersted accidentally discovered that an electrical current produced a magnetic field. It was later found that a changing magnetic field can create an electric current — an effect known as electromagnetism. This meant that, for the first time, a constant source of electricity could easily be produced. It was this discovery of the relationship between magnetism and electricity that made possible the next step in telecommunications.

MAKE AN ELECTROMAGNET

For safety reasons, ask an adult to help you make this electromagnet. You will need the following (available from good hardware stores):

a 1.5-volt battery, an on/off switch, coated wire, a large iron nail, copper wire, some paper clips, tape

1. Wrap the copper wire tightly around the nail, keeping the coils very close together.
2. Use some tape to make sure the wire stays tight.
3. Cut three pieces of coated wire. Ask an adult to cut back enough plastic coating to expose about 1 in. (2.5 cm) of wire at each end.
4. Connect one piece from the battery to one end of the copper coil; one piece from the battery to the switch and the third from the switch to the other end of the coil to make a circuit.
5. Turn the switch to "on" and the nail will be magnetized. Use it to pick up some paper clips.
6. Now turn the switch off. What happens to the paper clips?

WAVES THROUGH THE AIR

In 1887, the German scientist Heinrich Hertz demonstrated that electromagnetic power was not restricted merely to traveling along wires, but could exist as waves that move through the air. It was this discovery that eventually led to the invention of the radio.

THE FIRST BATTERY

In 1801, the Italian scientist Alessandro Volta demonstrated his battery to Napoleon Bonaparte. Napoleon was so impressed that he made Volta a count.

VOLTA'S PILE

The invention of the battery in 1800 by Alessandro Volta meant that for the first time electricity could be stored and released. It worked by stacking discs made of different metals on top of one another, each separated by a piece of felt soaked in acid. The chemical reaction between the different metals produced a weak electric current. In the 1830s, before electromagnetism became widely used to produce an electric current, the telegraph systems were powered by Volta's batteries.

FAST-MOVING ELECTRICITY

This picture shows a model of a famous experiment carried out by Benjamin Franklin that increased our understanding of electricity. Franklin wanted to know whether electricity could travel, and also if lightning was an example of this. During a thunderstorm in 1752, he flew a kite with a metal tip. There were two lines coming off the kite. The one he held was silk, a poor conductor of electricity, while the other was hemp, a good conductor. The hemp line had a metal key attached to the end. The experiment was very dangerous but Franklin proved his theory when he drew sparks from the key.

USING A TEMPORARY MAGNET

This picture shows one of the more obvious uses of electromagnetism. An electrical current passes through the metal disc and makes it magnetic. When the current is switched off, the scrap metal drops off. Most electromagnets are made as coils of wire called solenoids. By winding the wire into many loops, a strong magnetic field is created.

MAGNETISM MAKES ELECTRICITY

In 1831, two scientists, Michael Faraday (right) and Joseph Henry, both discovered that a static magnetic field does not produce an electrical current. However, a changing magnetic field does when passed through a coil of wire. This important discovery led to the creation of machines that could generate electricity.

WIRES UNDER THE SEA

In July 1866, the *Great Eastern*, a huge ship designed by British engineer Isambard Kingdom Brunel, set off across the Atlantic with 2,770 miles (4,457 km) of cable and successfully connected Britain and America. By 1880, there were nine cables across the Atlantic, which, if put end to end, would have measured 97,500 miles (156,903 km).

MAKE A MORSE CODE SENDER

Instead of sounds or printed dots and dashes, this very simple machine uses a lightbulb to send messages. You will need the following (available at good hardware stores):

an on/off switch, a 1.5-volt battery, a small lightbulb and bulb holder, coated wire, scissors

1. Cut three pieces of the coated wire (each about 12 in/30 cm long) and ask an adult to strip the plastic coating from each end for you so that about 1 in. (2.5 cm) of wire is exposed.
2. Take a wire and attach one end to the battery and the other to the on/off switch.
3. Take another wire and attach that from the on/off switch to the bulb holder.
4. Attach the last piece of wire from the other end of the battery to the lightbulb holder. This completes the circuit. Now you can send messages in Morse code by using the on/off switch. A short burst of light represents a dot and a longer burst a dash.

WIRES OVERHEAD

This picture shows the confusion of wires that became a feature of many city streets as the use of the telegraph expanded. Underground cables soon got rid of this unsightly mess.

THE MORSE CODE

Morse invented his code in 1838. Messages were initially printed out. A short electrical current was printed as a dot and a longer electrical current as a dash. He later realized that messages could be read more easily if the dots and dashes were replaced by clicking sounds of different lengths to distinguish dots from dashes.

A	• —	N	— •
B	— • • •	O	— — —
C	— • — •	P	• — — •
D	— • •	Q	— — • —
E	•	R	• — •
F	• • — •	S	• • •
G	— — •	T	—
H	• • • •	U	• • —
I	• •	V	• • • —
J	• — — —	W	• — —
K	— • —	X	— • • —
L	• — • •	Y	— • — —
M	— —	Z	— — • •

10

THE ARRIVAL OF THE TELEGRAPH

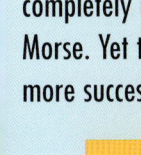

Throughout the history of electrical telecommunications, there is still some controversy about who discovered or invented what. The people who were working in this field were mainly enthusiastic amateurs rather than trained scientists. Many realized that it might be possible to send messages using electrical currents along a wire. In 1753, the German Thomas von Sömmering devised a telegraph system using electrolysis, a process which uses electricity to split water into hydrogen and oxygen. Transmitted signals appeared as bubbles of gas from wire terminals in a water tank. Eighty years later, a primitive form of telegraph powered by a voltaic pile was created by two Germans named Carl Gauss and Wilhelm Weber. Lack of money, however, meant that their efforts were completely ignored. For many, the inventor of the telegraph was Samuel Morse. Yet this is not really the case — it is simply that Morse's ideas were more successful than those of his predecessors.

MORSE & HIS MACHINE

In 1843, Morse got the support of the American Congress to build a telegraph line from Washington to Baltimore, a distance of 40 miles (64 km). The first message was tapped out on May 24, 1844. It was "What hath God Wrought."

THE FIVE-NEEDLE TELEGRAPH

In 1837, the five-needle telegraph was invented in Britain by Charles Wheatstone and William Cooke. It worked by sending an electric current along wires that moved two of the five needles either left or right so that they both pointed to one letter at a time.

COMMUNICATING BETWEEN STATIONS

The telegraph flourished because it came along at the same time as the railway building boom, which needed a fast form of communication between stations. In 1839, the engineer Isambard Kingdom Brunel convinced British railway companies to use Cooke and Wheatstone's telegraph. American railway companies used Morse's system instead.

THE ARRIVAL OF THE TELEPHONE

The rise of the telegraph allowed people around the world to be able to send messages to each other in an instant. Yet the telegraph was only the beginning. Within just four decades, it would begin to give way to an invention that carried the human voice over vast distances. Most people think that the Scottish-born Alexander Graham Bell was the inventor of the telephone. But many of his ideas came from a German teacher named Philipp Reis who invented a simple telephone in 1860, 16 years before Bell. Reis died after having built just a dozen of his telephones. One of these reached Edinburgh University where Bell was a student. It inspired him to try and build his own telephone after he moved to Boston, Massachusetts, in 1871. In America he was not the only person trying to invent a successful telephone. Elisha Gray and Thomas Edison were also searching for a way to transmit the voice.

NEW DISCOVERIES

The rise of the telephone came at the same time as the discovery of a new material called plastic. The first plastic was called Bakelite and allowed telephones to be created in many different shapes, including this "daffodil" telephone.

THE FIRST TELEPHONE MESSAGE

Bell first spoke into his newly invented telephone in 1876. He spoke from his attic to his assistant, Thomas Watson, who was waiting in another room. Bell said, "Mr Watson, please come here, I need you!" Watson rushed upstairs with the news of the first telephone transmission. Little more than a decade later, the use of the telephone had become widespread. This picture shows Bell opening the New York to Chicago telephone line in 1892.

EXPERIMENTING WITH MESSAGES

Bell experimented for many years with different ways of sending and receiving spoken messages. This picture shows one of the earliest machines he built for sending messages. It is called the Gallows-Frame transmitter.

A COMPETITOR AND IMPROVER

The great American inventor Thomas Edison was also working on creating a telephone but Bell beat him to it. However, in 1878, Edison invented a microphone that considerably improved on Bell's telephone. The microphone made the voice of the person speaking much clearer to the listener.

FIGHTING FOR BUSINESS

Bell obtained his patent for the telephone in 1876. From then on, Bell's telephone company was constantly battling with others who wanted to use his invention. Werner Siemens discovered that Bell had forgotten to patent his invention in Germany and by 1877 was mass-producing an improved version of Bell's telephone there. The picture shows an advertisement for the public telephone system set up by the Bell Telephone Company.

THE RISE OF THE TELEPHONE

Perhaps because the telegraph had only recently been established, there was not much interest in Bell's creation at first. Many people thought that it was little more than a toy. In Britain, W.H. Preece, chief engineer for the Post Office, said that only Americans needed telephones because they lacked servants to carry messages for them. Bell overcame the problem by continually demonstrating his telephone and by employing the world's first public relations advisor. On January 14, 1878, he showed his invention to Queen Victoria. She was impressed, and decided to have a telephone installed in one of her houses. Once the initial doubts about the telephone were overcome, its use expanded rapidly at the telegraph's expense. By 1887, there were over 100,000 telephones around the world, mostly belonging to wealthy Americans and Europeans.

SENDING PICTURES BY TELEPHONE

The first fax service was opened in France in 1865 and was used to send photographs to newspapers. Each fax page is divided into tiny squares called Pels which are either black or white. These are sent along the telephone wire like the dots and dashes of Morse code and put together at the other end.

USING THE OPERATOR

The story of the invention of direct dialing in 1889 is a strange one. In Kansas City, an undertaker named Almon Strowger discovered that his local operator was married to a rival undertaker and was diverting his calls to her husband. This drove Strowger to find a way to cut out the operator.

MODERN DESIGNS

In today's cities, telephone booths are designed not so much for privacy as for keeping out traffic noise through the use of soundproofing.

TELEPHONE BOOTHS

In 1889, the first coin-operated telephone booth was opened in the U.S. Many countries began to copy the idea and soon telephone booths were seen in many towns and cities. Perhaps the most famous example is the British red telephone booth, designed by Sir Giles Gilbert in 1936. This type of booth was so popular that its replacement in the 1980s caused a public outcry.

THE COAXIAL CABLE

As the telephone became more popular, ordinary cables could not cope with the demand. Coaxial cables were first used in England in 1936. Each of these cables can carry hundreds of signals at a time. They are still in use today.

LOOKING INSIDE A TELEPHONE

1. The earpiece: The loudspeaker inside the earpiece has a coil of wire around a magnet. An alternating current makes this coil into an electromagnet. The two magnets constantly repel and attract each other, which makes the coil move. This movement vibrates a thin cone that then recreates the original sound.

2. The cable: The electrical signal passes down the cable to the local telephone exchange.

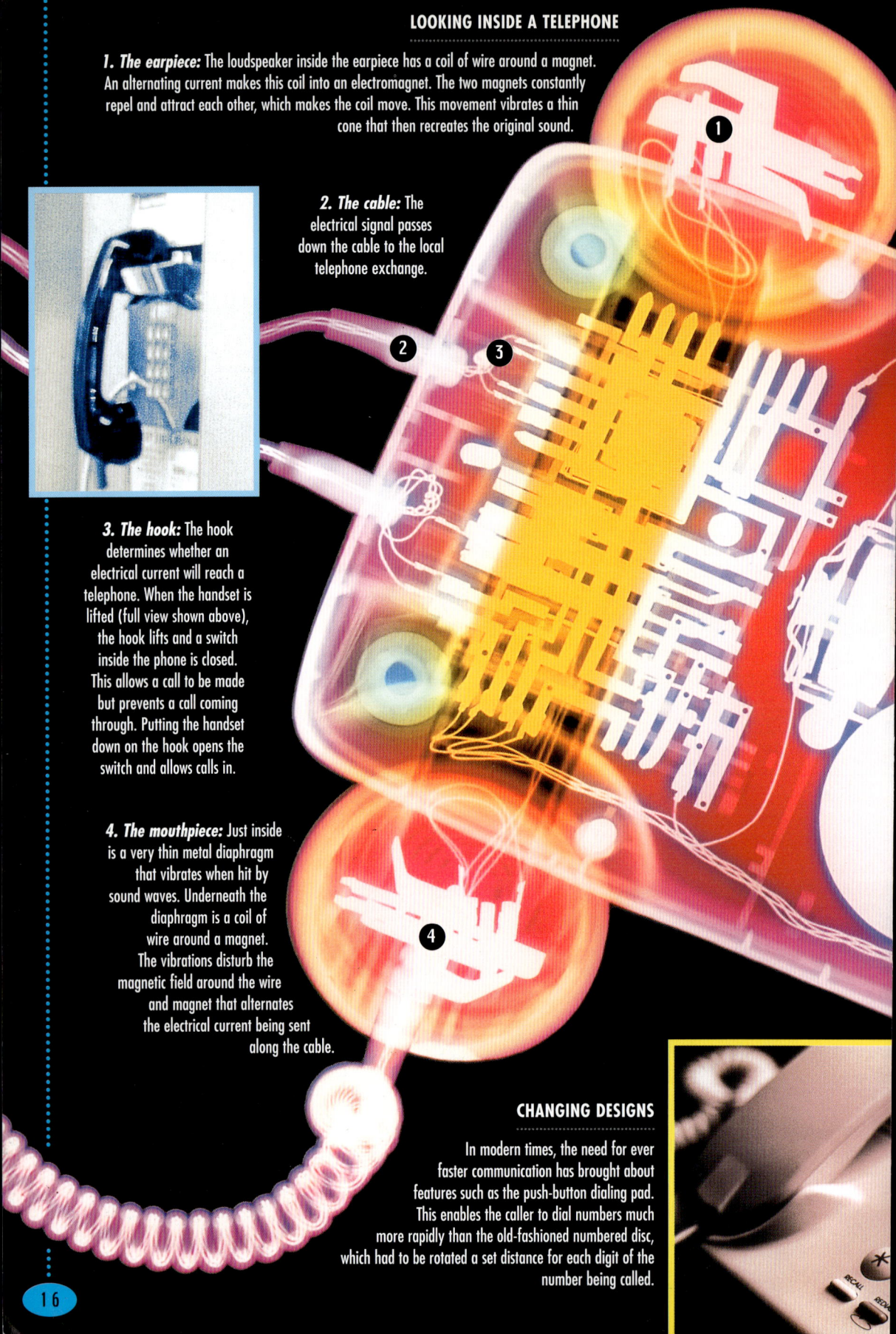

3. The hook: The hook determines whether an electrical current will reach a telephone. When the handset is lifted (full view shown above), the hook lifts and a switch inside the phone is closed. This allows a call to be made but prevents a call coming through. Putting the handset down on the hook opens the switch and allows calls in.

4. The mouthpiece: Just inside is a very thin metal diaphragm that vibrates when hit by sound waves. Underneath the diaphragm is a coil of wire around a magnet. The vibrations disturb the magnetic field around the wire and magnet that alternates the electrical current being sent along the cable.

CHANGING DESIGNS

In modern times, the need for ever faster communication has brought about features such as the push-button dialing pad. This enables the caller to dial numbers much more rapidly than the old-fashioned numbered disc, which had to be rotated a set distance for each digit of the number being called.

16

HOW THE TELEPHONE WORKS

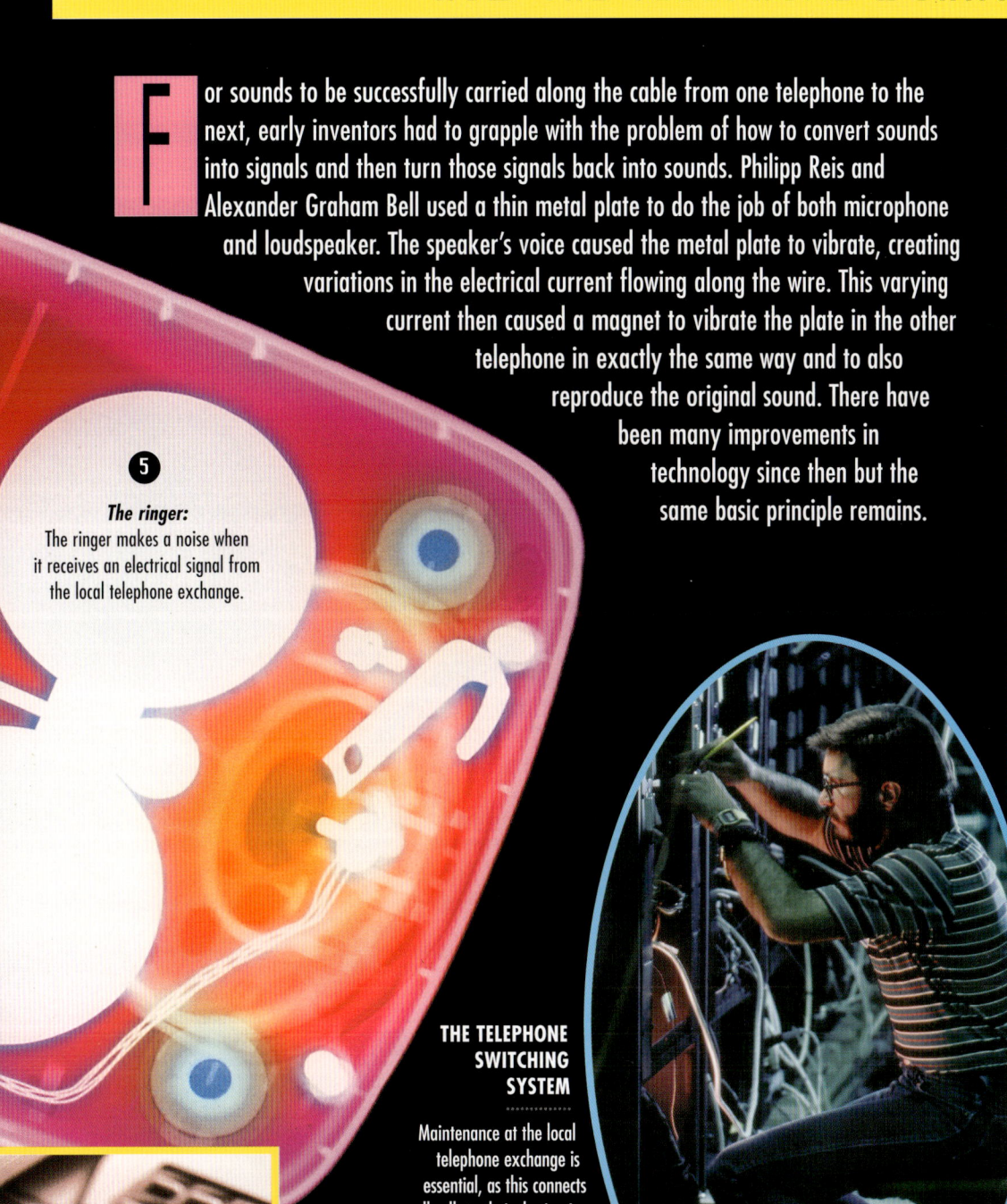

For sounds to be successfully carried along the cable from one telephone to the next, early inventors had to grapple with the problem of how to convert sounds into signals and then turn those signals back into sounds. Philipp Reis and Alexander Graham Bell used a thin metal plate to do the job of both microphone and loudspeaker. The speaker's voice caused the metal plate to vibrate, creating variations in the electrical current flowing along the wire. This varying current then caused a magnet to vibrate the plate in the other telephone in exactly the same way and to also reproduce the original sound. There have been many improvements in technology since then but the same basic principle remains.

⑤
The ringer:
The ringer makes a noise when it receives an electrical signal from the local telephone exchange.

THE TELEPHONE SWITCHING SYSTEM

Maintenance at the local telephone exchange is essential, as this connects all calls to their destination. Local exchanges can only directly connect local calls. For a long-distance call, the signal goes from the local exchange to an area exchange. It is then routed through several area exchanges, then to the nearest local exchange and finally to the telephone being called.

THE DISCOVERY OF RADIO

All of these advances in telecommunications took place in a remarkably short space of time. Yet they were still restricted by the use of cables. If there were no cables connecting the caller and the receiver they could not communicate with each other. In 1867, the Scottish scientist James Clerk Maxwell discovered that changes in an electromagnetic field happened at the speed of light. He thought that perhaps there were different kinds and sizes of electromagnetic waves that traveled through the air and that light was one of those waves. His theory was right but he was unable to prove it. In 1888, the German scientist Heinrich Hertz discovered that electromagnetic waves did travel through the air. The waves that he found were much longer than waves of light and became known as radio waves.

GUGLIELMO MARCONI

It was the Italian scientist Guglielmo Marconi who realized that the newly discovered waves could transmit messages. In 1894, he made a bell ring by transmitting electromagnetic waves toward it. He moved to Britain in 1896, where he worked on developing long-distance radio signals.

POLITICAL BROADCASTS

The first radio transmission of human speech was made in 1906 by Reginald Fessenden, broadcasting from Massachusetts to ships in the Atlantic. Fessenden had found a way of varying radio waves to mimic sound waves. Politicians, such as Winston Churchill (above), were quick to recognize the power of radio, and used it to broadcast many of their speeches.

MARTIANS ON AIR

In 1938, the Mercury Theater broadcasted a live version of H.G. Wells' *The War of the Worlds*. Many listeners panicked, as they believed it was a report of a real Martian invasion. Such an incident seems less surprising when you consider that, before the arrival of television, radio provided the main link with the outside world for many listeners. Unlike newspapers, radio was able to bring them up-to-the-minute information on the day's events, just as it continues to do today.

IMPROVING SOUND QUALITY

The triode valve (left) was invented by Lee De Forest in 1906. It amplified radio signals that improved the quality of the sound created by radio waves.

FROM ENGLAND TO CANADA

Until 1906, all radio messages were in Morse code. This picture shows Marconi's wireless telegraph. Marconi became a household name in 1901 when he broadcast a radio signal in Morse code from Cornwall across the Atlantic to Newfoundland. This was the first transatlantic radio communication.

CAPTURED BY RADIO

Radio saved many of the 700 survivors of the *Titanic* in 1912 when the SOS signal was sent out as the ship sank. In 1910, radio also helped to catch the murderer Dr. Crippen, who was escaping from England to Canada. Crippen was recognized, a message was sent to the Canadian police, and he was arrested on arrival.

HOW RADIO WORKS

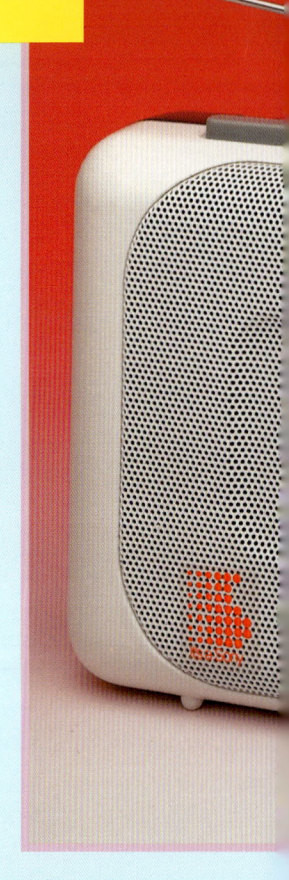

The arrival of radio stations broadcasting speech and music to people's homes was delayed by World War I. The first town to have its own permanent radio station was Pittsburgh, Pennsylvania, which started broadcasting in November 1920. The British were close behind when the British Broadcasting Corporation (BBC) was founded in 1922 and set up radio stations in London, Birmingham, and Manchester. After World War II, the valve began to be replaced by the transistor, which had been invented in 1948. By the 1970s, the valve had more or less disappeared. The transistor meant that, for the first time, cheap and portable radios could be easily made. It also led to the first car radios being built. Radio broadcasting has changed with the rise of television, but it is still the most important form of telecommunication in the world.

THE TRANSMITTER

Radio transmitters work by sending out radio waves from an antenna. Electrical signals set at a particular frequency are mixed with the electrical signals created by sounds to be broadcast. These are called modulated signals and produce the radio waves that are sent out in all directions.

LISTENING AROUND THE WORLD

In 1991, British inventor Trevor Baylis created the wind-up radio, enabling millions in the developing world, with no permanent electricity supply, to receive broadcasts. The radio works by winding up a spring, which slowly uncoils and powers a small internal generator. The radio is produced in South Africa and over 20,000 a month are being made.

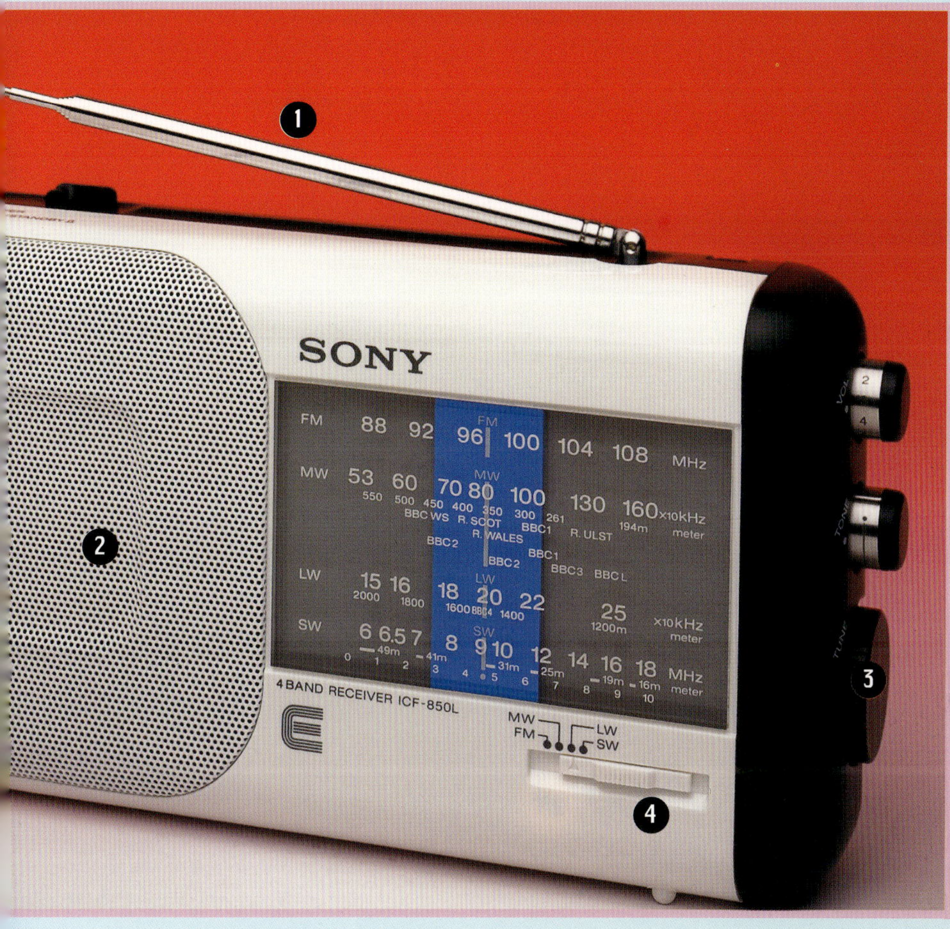

PARTS OF A RADIO

1. The antenna: The antenna receives radio waves that are carrying a sound signal. The radio waves produce a changing electrical current in the antenna, which is then sent into the radio.

2. The loudspeaker: The radio separates the radio signal from the radio wave and sends it to the loudspeaker.

3. The tuning knob: Each radio station broadcasts on just one radio frequency. The tuning knob can select only one frequency at a time.

4. The waveband selector: Frequencies are divided into groups called wavebands. The waveband selector allows frequencies to be searched within each waveband.

DIFFERENT WAVES

All of the different wavebands behave in different ways. This means that they can be used for different kinds of radio broadcasting.

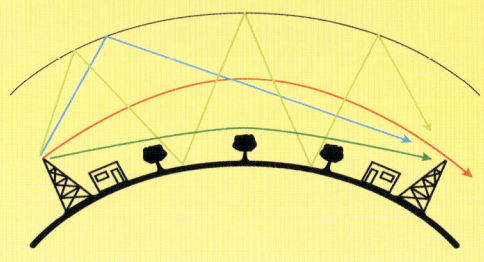

Earth's atmosphere (ionosphere)

Short waves can bounce off the ground and the ionosphere, a layer in the Earth's atmosphere. This allows short wave radio stations to be heard all over the world.

Medium waves can bounce off the ionosphere, but not the ground, so they have a smaller range than short waves.

Long waves get their name because they have a low frequency, which gives them a long wave length.

VHF (very high frequency) waves cannot reflect off the ionosphere, but can only travel in straight lines or bounce off the ground. This means that they cannot travel very far and are normally used for local radio.

CARRYING THE SOUND – AM & FM

Making radio waves carry sounds is called modulation. There are two ways in which this can happen, amplitude modulation (AM) and frequency modulation (FM). In AM, the different sounds are carried by changing the modulation (or height) of the radio wave. In FM, the sounds are carried by varying the frequency of the waves.

— Carrier radio wave   — Sound wave

AMPLITUDE MODULATION

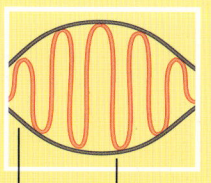

Low amplitude   High amplitude

FREQUENCY MODULATION

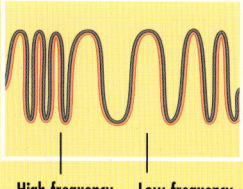

High frequency   Low frequency

21

HOW TELEVISION WORKS

The idea of transmitting pictures by electricity occurred at around the same time as the invention of the telegraph. The fax machine was able to transmit still pictures, and was in common use after 1865, especially by the news agencies such as Reuters. However, the problem of how to send moving pictures across wires began to be solved in 1873. It was found that the element selenium resisted an electric current in the dark but let through more current when it became lighter. Ten years later, a German named Paul Nipkow used this discovery to invent the Electrical Telescope – what we call television. Nipkow realized that selenium could be used to produce a picture that could be divided into very small points. These points could then be carried on radio waves and reassembled at the other end. Finally, in 1897, J.J. Thomson discovered the electron, and this led to the invention of the cathode-ray tube used to create pictures on television screens.

CREATING PICTURES

A close look at a picture on a television screen will show that it is made up of hundreds of horizontal lines. These in turn are each made up of hundreds of multi-colored dots. The lines and dots merge together to create the illusion of a complete picture. The movement of television pictures is also an illusion, created by showing lots of still pictures in rapid succession. Each picture is slightly different from the one before. The eye and the brain of the viewer is fooled into seeing a moving picture – a process known as persistence of vision.

USING THE REMOTE

Remote controls work by sending out an infrared signal to a photodiode on the television. A photodiode changes an infrared signal into an electrical one. This new signal then changes the channel on the television.

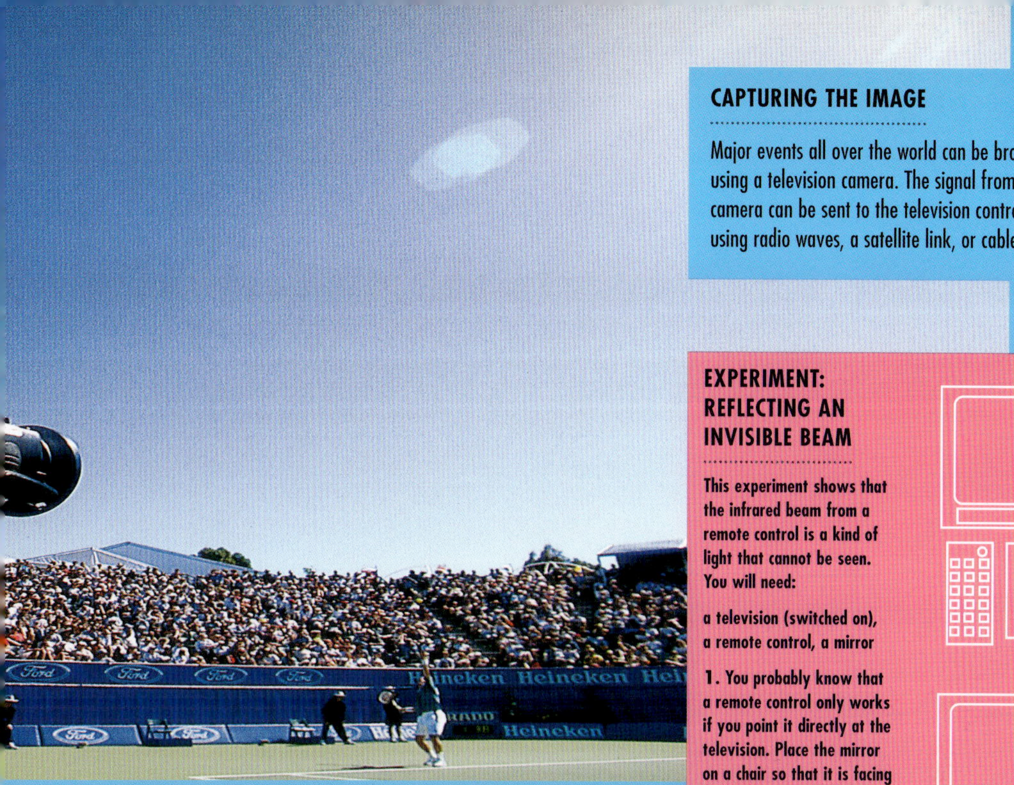

CAPTURING THE IMAGE

Major events all over the world can be broadcast using a television camera. The signal from the camera can be sent to the television control room using radio waves, a satellite link, or cable.

EXPERIMENT: REFLECTING AN INVISIBLE BEAM

This experiment shows that the infrared beam from a remote control is a kind of light that cannot be seen. You will need:

a television (switched on), a remote control, a mirror

1. You probably know that a remote control only works if you point it directly at the television. Place the mirror on a chair so that it is facing the television.
2. Sit on the floor with your back to the television. Point the remote towards the mirror and you can change channels. If it does not work, try moving the mirror until it does. Make sure you are not sitting between the mirror and the television. This works because light bounces off a mirror.

SHAPING PROGRAMS

Inside the television control room producers can choose how programs are put together before they are broadcast.

INVENTING THE TELEVISION

It was John Logie Baird who invented the first practical television in 1926. He used a mechanical camera with a spinning disc in front. This produced a strobe-like effect, giving the appearance of motion, but the quality of the images was not very high.

23

THE DIFFERENCE ENGINE

The first computer was designed in 1832 by Charles Babbage and was called the Difference Engine. It was capable of doing complex mathematical equations, but it used only simple addition for its calculations. There was no need for multiplication or division. Before this, calculations were done using mathematical tables by people called "computers."

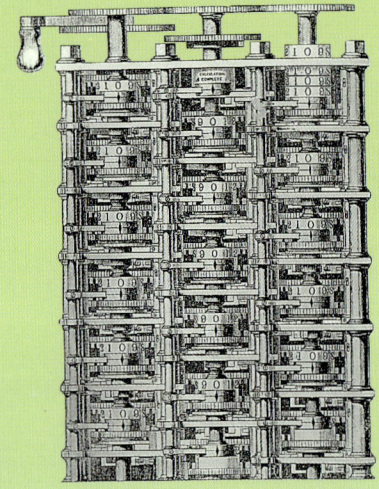

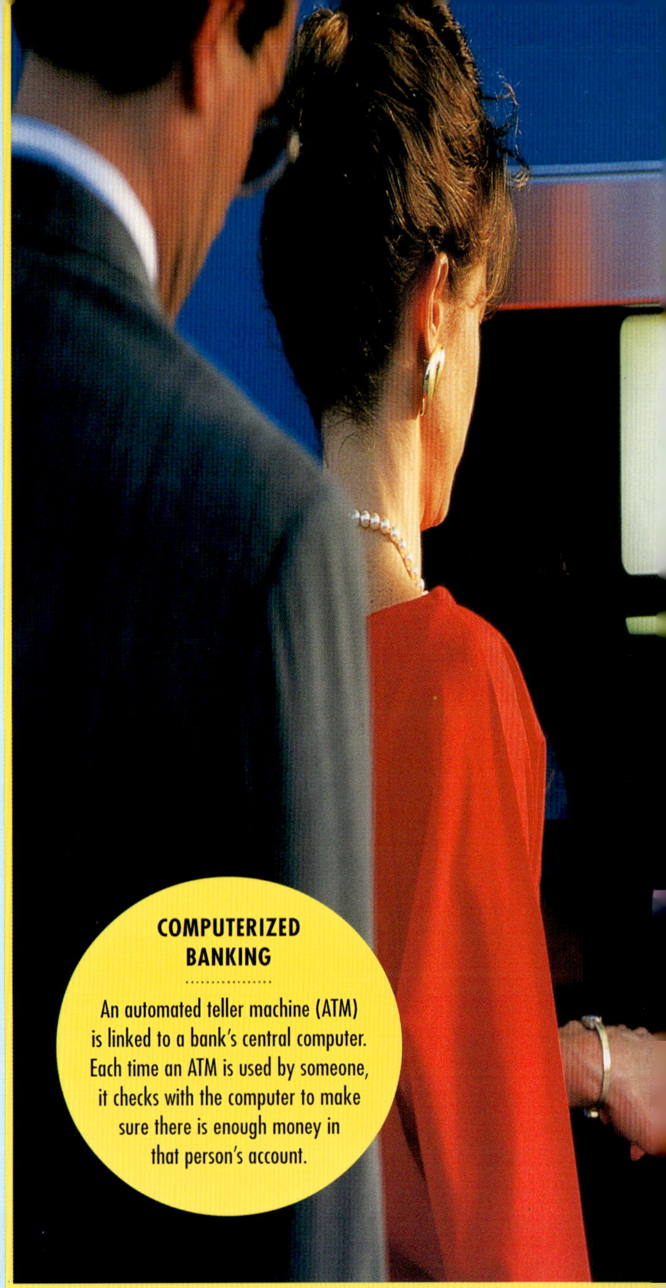

THE INTERNET

The fastest growing form of telecommunication is the Internet. Messages can be sent by E-mail and information in the form of text, pictures, sound, and video can be found on the World Wide Web. The Internet grew from an American military computer network that was set up in the 1960s.

COMPUTERIZED BANKING

An automated teller machine (ATM) is linked to a bank's central computer. Each time an ATM is used by someone, it checks with the computer to make sure there is enough money in that person's account.

VIDEOCONFERENCING

Many companies can now hold meetings without needing to have everyone in the same room. Videoconferencing works with a microphone and a video camera attached to a computer. Sounds and pictures are converted into digital form and are then sent to the other participants' computers. Although the transmission travels at the speed of light, videoconferencing is still not quite "real time." There is a slight delay because of the need to turn information into a digital form and then convert it back again.

COMPUTER COMMUNICATION

All of us are using computers nearly every day of our lives. Even if people do not use a computer at home, school, or work, they are still communicating with computers, often without realizing it. Every time people go shopping in a supermarket, get money from a bank, or borrow a book from a library, communication with a computer is taking place. What is remarkable about these supposedly complex machines is that all the information they receive, store, and send is in the form of digital information using only the numbers 0 and 1. Together, these numbers make up a system of counting known as binary. A computer creates a 1 by turning on the current and a 0 by turning off the current. All of the information on a computer, whether words, pictures, sound, or video, is nothing more than huge combinations of 0s and 1s.

NETWORKING IN THE OFFICE

Most people in offices communicate through a network of computers. All of the computers in an office or building are linked through a local area network or LAN. Computers in different buildings are linked together through a wide area network or WAN.

SATELLITE & DIGITAL TECHNOLOGY

Many of the commercial breakthroughs in telecommunications have happened because they were supported by a growing system of transport. The telegraph was helped by the development of railways. Radio was first enthusiastically taken up by the shipping companies. In the same way, truly worldwide communication through the telephone, computer, or television would not have been possible without the advent of space travel. In October 1957, the Soviet Union launched the *Sputnik* satellite, which sent a radio signal back to Earth. The "Space Race" had begun. In the summer of 1962, America launched the *Telstar* satellite. *Telstar* provided a radio and television link between Europe and America for just a few hours a day. Within just three years a new type of satellite named *Early Bird* was blasted into space and, for the first time, created the reality of instant and constant worldwide communication.

BEING MOBILE

Mobile phones use radio to communicate. Geographical areas are divided into cells (see diagram below). In each cell is an antenna that keeps track of all the mobile phones within that cell. This is how a telephone message gets through to the right mobile phone. In remote areas such as at sea or in the desert, communications satellites orbiting the Earth do the same job as the antenna.

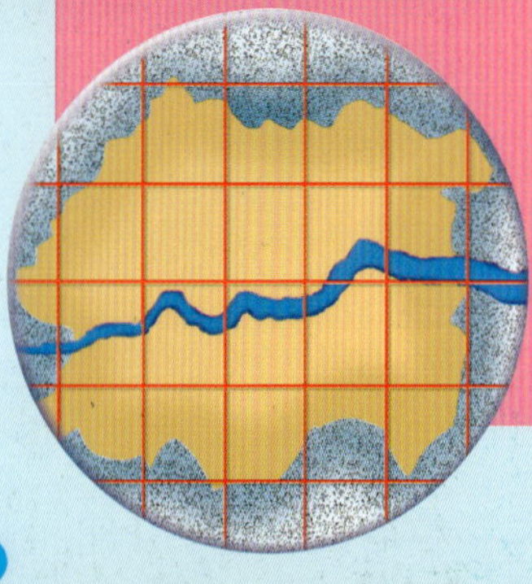

COMMUNICATING IN SPACE

In space, communication takes place through the use of radio waves. Astronauts talk to each other and contact Earth using radio. Unmanned satellites are controlled by radio waves and send back information in the same way. Radio waves travel at the speed of light, but the signals from satellites that have traveled past distant planets such as Pluto can take several hours to reach Earth.

FOCUSING THE WAVES

The radio waves from an antenna radiate in all directions. These waves bounce off the satellite dish in the same direction. Any radio waves coming in hit the dish and bounce onto the antenna in the middle.

COMMUNICATION TOWER

This British Telecom Tower in London has dishes that communicate with satellites using microwaves. These waves can be sent into space as a much narrower beam than other types of electromagnetic wave, particularly radio waves. This allows transmission to be more precise and have less interference. Communications satellites are set to exactly match the rotation of the Earth, so they appear to be standing still. They are known as "geostationary" satellites.

DIGITAL & ANALOG BROADCASTING

These two pictures below show how digital broadcasting works. Traditionally, sounds and pictures are broadcast in an analog form. This means that sounds and pictures are converted into waves and then transmitted. Digital broadcasting uses the same technology as computers. Sounds and pictures are converted into binary (0s and 1s), which is then transmitted through cable, satellite, or transmitters.

Analog Transmission

Digital Transmission

DIGITAL RADIO

Although the area of digital television is attracting most interest, digital radio will also bring many improvements. Because it uses the same technology as compact discs, digital radio can produce CD-quality sound. Interference will be reduced and there will be no need to re-tune car radios when traveling from one part of the country to another.

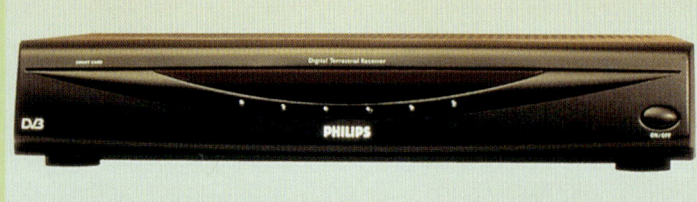

DIGITAL DECODER

As digital television becomes more common, television sets that can read these new signals will be built. It is possible for today's televisions to receive digital signals as long as they have a digital decoder (above). This converts the digital signal into an electrical signal that the television can read.

EXPERIMENT: HOW FIBER-OPTICS WORK

Fiber-optics work because light travels in a straight line. When it hits an obstacle, it is either reflected outward or refracted inward.
This experiment gives an approximate idea of how light bounces along inside a fiber-optical cable.

You will need:

a large cardboard box, a flashlight, several small, flat mirrors, black crayon or paint, a darkened room

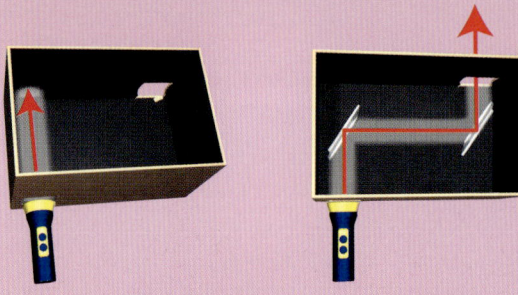

1. Make the whole of the inside of the box black, using paint or crayons.
2. Cut a hole next to one of the bottom corners of the box so that it is just big enough for the flashlight to fit through. Cut another hole diagonally opposite to this first hole.
3. In a darkened room, put the flashlight through the first hole and switch it on. The light travels in a straight line and hits the opposite side of the box.
4. Now try to get the light to shine through the other hole. Do this by positioning a mirror in front of the beam of light so that the light bounces off. You can then use more mirrors to bounce the light through the vacant hole. With practice, you should be able to do it using only two mirrors.

SATELLITE & DIGITAL TECHNOLOGY – THE FUTURE

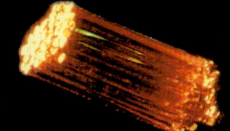

The huge rise in radio telecommunications is beginning to cause a problem. As the number of mobile telephones and radio and television stations continually grows around the world, the amount of space available on the different wavebands is shrinking. This means that there is more risk of interference from other broadcasters and transmission quality goes down. The weather and other atmospheric changes also affect the quality of broadcasting. Much of this will change with the arrival of digital telecommunications. Digital signals can be compressed into a much smaller space than traditional methods of broadcasting. It also means that the quality of sounds and pictures will greatly improve since the signal will not be affected by the weather. It will allow people to view the World Wide Web on their televisions and for television programs to become truly interactive by integrating the Web with television.

COMMUNICATING WITH LIGHT

Fiber-optical cables are an important part of cable digital communication. Inside each cable are hair-thin tubes of glass protected by a plastic layer. A laser beam is sent along the fiber and is changed into pulses of light by an electrical signal. These pulses can then be converted back into electrical signals.

TELECOMMUNICATIONS AT WORK & HOME

Rapid changes in modern telecommunications mean that there are now very few areas of our personal and working lives that have not changed beyond recognition. There are many benefits, such as reducing travel to workplaces, since it is now possible to telecommute. This means that people can work from home and link up to their office through their computer. Employees and employers can live on different continents, since distance is irrelevant. There are some, however, who feel that all of these rapid changes may actually be harming both individuals and the communities in which they live since it is no longer necessary to meet people face-to-face. Yet the vast amount of information that modern telecommunications readily provides — from global news to educational programs — means that people are now better informed about the world than at any other time in history.

ALL OVER THE WORLD

Even in the frozen wastes of the North Pole it is possible to keep in touch. Since mobile telephones and the Internet can communicate through satellites in space there is now nowhere on Earth that cannot be reached through telecommunications.

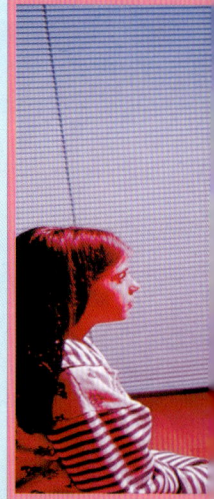

COMPUTERS IN THE CLASSROOM

Learning in the classroom no longer relies solely on books. Children now find information on CD-ROMs and the Internet, as well as in books.

WATCHING THE WORLD

This picture shows President Kennedy talking to the American people during the Cuban missile crisis in 1962. Satellite television means that nearly every major event can now be seen as it happens. The fall of the Berlin Wall, the Gulf War, and the conflict in Kosovo became real for many through the power of television.

KEEPING IN TOUCH

The rise in the use of mobile telephones means that people can now be contacted at any time and in virtually any location. It is estimated that in Britain, the number of people using mobile telephones is rising by about 10,000 a week.

TOO MUCH OF A GOOD THING?

There are some worries about the amount of time children spend watching television. In the U.S., teenagers spend about 900 hours per year in high school but watch more than 1,500 hours of television per year. Watching too much television has been linked to aggressive behavior and low academic achievement. The rise in digital, satellite, and cable television may add to the problem as it has led to a massive rise in the number of television channels available. It will soon be possible to choose between several hundred television channels.

GLOSSARY

Analog broadcasting
In analog broadcasting, sounds and pictures are converted into waves and then transmitted.

Digital broadcasting
In digital broadcasting, sounds and pictures are converted into binary (0s and 1s), which is then transmitted through cable, satellites, or transmitters.

Electromagnetism
Electricity and magnetism are very closely linked. An electric current will always produce a magnetic field. A changing magnetic field always creates an electrical current, and this current operates many of the instruments used in modern telecommunications.

Electrons
Electrons are tiny particles that orbit the nuclei of an atom. Some of the electrons that revolve around metal atoms can escape. It is these electrons that can carry an electric current along a circuit.

Fiber Optics
Rays of light can pass along a thin glass thread called an optical fiber. An outer coating of a different kind of glass reflects the light inwards so that it can never escape from inside the fiber. Optical fibers are used to carry signals along telephone cables.

Frequency
Radio waves have different frequencies. Frequency is measured by the number of times something repeats itself every second. In the case of radio waves, frequency means the number of wave crests that go over a particular point every second.

Satellite broadcasting
Using a special satellite dish antenna, homes can receive television or radio programs beamed directly from broadcasting satellites orbiting the Earth.

World Wide Web
The World Wide Web, or Internet, is a huge network of information that can be accessed by computer. A piece of equipment called a modem is needed to connect a computer to the Internet.

Wavelength
This is the distance between the tops of two waves when the radio wave is regular.

ACKNOWLEDGMENTS

We would like to thank David Rooney and Elizabeth Wiggans for their assistance. Artwork by John Alston.
First edition for the United States, its territories, dependencies, Canada, and the Philippine Republic, published 2000 by Barron's Educational Series, Inc.
Original edition copyright © 2000 by ticktock Publishing, Ltd.   U.S. edition copyright © 2000 by Barron's Educational Series, Inc.
All rights reserved. No part of this book may be reproduced in any form, by photostat, microfilm, xerography, or any other means, or incorporated into any information retrieval system, electronic or mechanical, without the written permission of the copyright owner.
All inquiries should be addressed to: Barron's Educational Series, Inc., 250 Wireless Boulevard, Hauppauge, NY 11788 • http://www.barronseduc.com
Library of Congress Catalog Card No. 99-67205   International Standard Book No. 0-7641-1068-3
9 8 7 6 5 4 3 2 1   Printed in Hong Kong   Picture research by Image Select.

Picture Credits: t = top, b = bottom, c = center, l = left, r= right, OFC = outside front cover, OBC = outside back cover, IFC = inside front cover

Allsport; 22/23t. Ann Ronan @ Image Select; 4c, 4cr, 5cl, 6br, 6bl, 8c, 8br, 9b, 10t, 11br, 11bl, 11cr, 12/13c, 18cl, 23br, 24tl. Corbis; 3cr, 3b, 4cl, 4/5t, 5cr, 6/7c, 7b, 10c, 15br, 17br, 18b, 18/19ct, 19bl, 30/31cb, 31r, 32tr. Image Select; 4/5c, 13cr, 19br, 20/21t, 23br. Nick Birch @ Image Select; 12bl, 13t, 13bl. Philips; 28c. PHOTRI/Image Select; 19tr. Pix S.A.; IFC , OFC (main), OFCbr, 2l, 3cl, 29 (main) & OFCtr. Rex Features London; 20b. Science Photo Library; 8/9t & OBC, 9r, 14cl, 16/17 (main, Hugh Turvey), 20l, 22/23cb, 24bl, 26/27b (David Ducros), 27r (Martin Bond), 30/31ct (Martin Fraser), 30/31c (Oscar Burriel). TCL Stock Directory; 2/3t. Telegraph Colour Library; 2br, 14b, 16tl, 16/17b, 22b, 25r, 26/27t, 30cl. Tony Stone Images; 15bl, 24/25t, 24/25b, 26l.

Every effort has been made to trace the copyright holders and we apologize in advance for any unintentional omissions.
We would be pleased to insert the appropriate acknowledgement in any subsequent edition of this publication.